Early Childhood Numeracy for the Caribbean

1 2 3 You and Me

Lorna Thompson

Advisor: Cathryn O'Sullivan

Activity Book 2

Introduction

Developed in accordance with curricula from across the region, *I, 2, 3 You and Me* is a complete package of interactive materials for Early Childhood Numeracy education in the Caribbean. It is child-centred, play-based, hands-on, developmentally appropriate and integrated. This package will help make teaching and learning enjoyable both at home and at school - everything you need to teach and excite young learners!

The *I, 2, 3 You and Me* **Activity Books** have been designed to promote Early Childhood Numeracy in a fun and interactive way. They can be used in conjunction with the full *I, 2, 3 You and Me* package, or can be used separately.

Activity Book I introduces foundational Numeracy skills through a range of activities that have a familiar sequence but become more challenging as the book progresses up to numeral I0.

Activity Book 2 builds on the skills that were introduced in Activity Book I by addressing the same material in a more complex manner and by introducing new concepts such as simple addition, as well as extending activities up to numeral I5.

Allow children beginning to develop their Numeracy skills to have fun and express themselves in these Activity Books - both books are designed to fit with and complement wider classroom or home activities, and encourage children to develop at their own pace. See how the teaching and learning of Numeracy can be enhanced with the full *I, 2, 3 You and Me* pack components, listed on the next page.

Material is ordered by numerals and interspersed with other Numeracy topics – each page has its own heading to outline what is being covered

'Looking back' pages are included throughout to help with the revision of topics as children progress

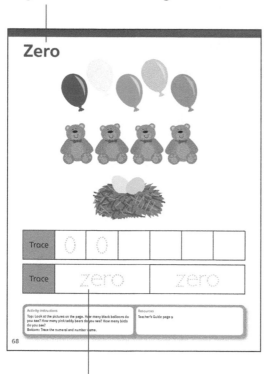

Each Activity Book includes lots of opportunity for children to draw and colour in

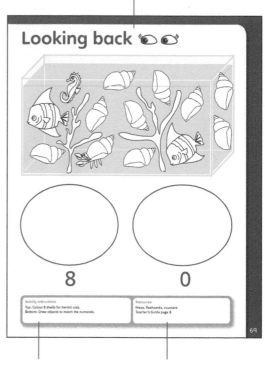

Pages introducing numbers include tracing activities with additional blank boxes for children to write the numerals if they are ready. If children are not ready to write, theses boxes can be used for additional tracing practice or be left blank

Activity instructions at the bottom of each page are there to guide the activity and should fit in with your approach to class/home discussions and other activities

Suggestions for incorporating hands-on activities using *I, 2, 3, You and Me* pack components, including the Teacher's Guide, are provided in the 'Resources' box at the bottom of each page. Be creative about using the materials

The *1, 2, 3, You and Me* package

Along with the **Activity Books**, here is what is included:

Teacher's Guide supports the full package with strategies to adapt regular classroom activities, including conversations to encourage critical thinking, activities for home and school, environmental observations, as well as a special focus on parental involvement and approaches to teaching children with special needs.

Parrot puppet is provided as a fun and engaging means of interacting with children. Its character is used throughout the books; use the puppet to introduce topics and with specific activities found in the Teacher's Guide as well as in the Activity Books. Let children give him or her a name!

Reward stickers - encourage and support children's efforts with reward stickers.

Flashcards are provided for numbers 0-12. Four sets of flashcards are provided to facilitate independent use as well as the playing of games like "go fish", "memory" or matching. The numeral and number name are on the front with quantities represented on the back. The arrangement of the pictures helps to introduce the concept of odd vs even, and the repeated background colours offer opportunity for colour recognition and matching games. Use the flashcards to complement the Activity Books, and learn more about additional ways to use them in the Teacher's Guide.

Numeral Frieze is brightly coloured and includes numerals 0-10. It's the perfect addition to the classroom wall. As with the flashcards, children can continue to learn about numerals, number words, colours and the concept of odd versus even.

Audio CD-ROM - have fun singing with your class with Numeracy-related songs and stories. Song lyrics and a numeral formation guide for the number formation song are provided in the Teacher's Guide.

Board Games - Four board games are included: (1)The Squeeze, (2) Beautiful Shapes, (3) Snakes and Ladders and (4) The Big Race. These promote the development of Numeracy skills, and as group games they encourage teamwork and sportsmanship. They can be used in conjunction with the dice, stickers and Unifix cubes. Detailed instructions for the board games, as well as suggestions for additional games, can be found in the Teacher's Guide.

Counter stickers are included for use with the games and activities including counting, sorting and matching. Use them on the top of everyday items like plastic bottle caps.

Unifix cubes - one set of 100 colourful, interlocking counting cubes represent 'units' and can be used in conjunction with the board games, Activity Book activities and additional games suggested in this Guide.

Dice with numerals and quantities are provided primarily for use with the board games. Children will also be able to start to associate the quantity with the numeral and can count the dots for verification. Suggestions for other activities involving dice can be found in the Teacher's Guide and the Activity Books.

Tangrams are an ancient Chinese puzzle comprised of seven shapes. Two sets of colourful tangrams are provided in this pack. They are versatile and can be put together in hundreds of ways to make a variety of shapes including people and animals. A detailed description of tangrams and how to use them is provided in the Teacher's Guide.

And everything can be stored in the reusable *1, 2, 3, You and Me* drawstring bag!

> Children are just beginning to develop Numeracy skills and it is important to encourage and foster their love of learning. Remember to praise them and their efforts and when checking for accuracy focus on the aim of the activity rather than their drawing skills.

Contents

Matching

Go together

Activity instructions

Draw a line between the two objects that go together. Talk about why they go together.

Resources

counters, cubes
Teacher's Guide page 7

Big and small

Activity instructions	Resources
Talk about the animals on the page. Circle the bigger animal in each pair. Put an X on the smaller animal.	parrot puppet Teacher's Guide page 14

3

· I one

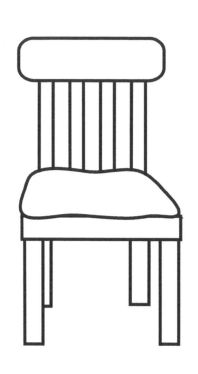

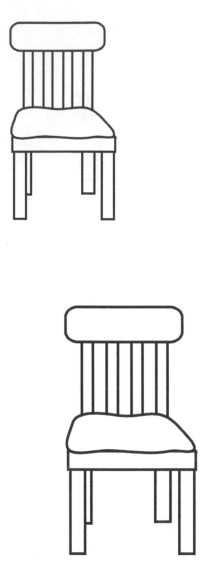

Trace	I	I					

Trace	one	one

Activity instructions

Top: Colour one chair for baby bear.

Bottom: Trace and write the numeral and the number name.

Resources

counters, Goldilocks story, songs: Number Formations;
Little Fingers
Teacher's Guide pages 8 and 9

What am I?

I can be found on all houses.

I have four sides.

Two long ones and two short ones.

I open and close. What am I?

Activity instructions

Draw an X on the correct answer. Colour the picture.

Resources

shape counters, Beautiful Shapes game, Tangrams
Teacher's Guide page 10

Same and different

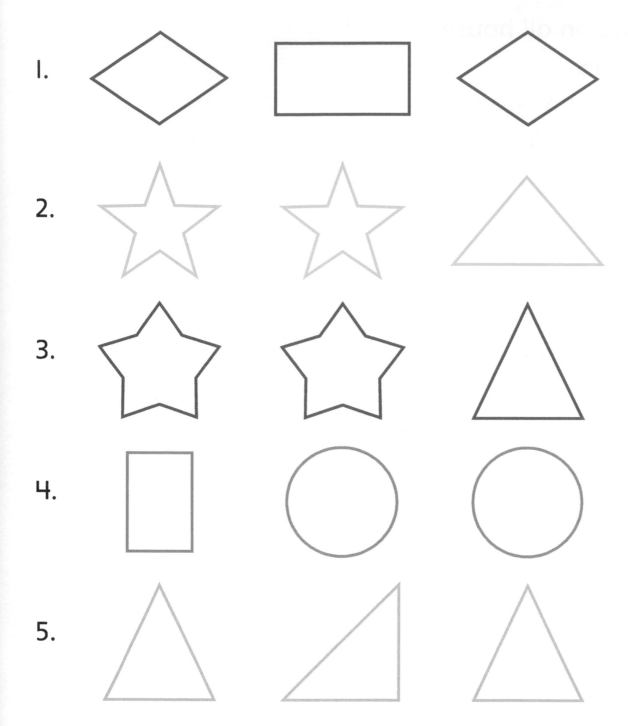

1.
2.
3.
4.
5.

Activity instructions

Talk about the shapes on each line. Colour the shapes that look the same on each line. [Talk about number 5 with the children and the fact that they are all triangles but they don't all look the same]

Resources

Beautiful Shapes game, shape counters, cubes, tangrams, Shape song
Teacher's Guide page 10

6

Same

Activity instructions

Draw lines to match the balls that are the same. Colour them.

Resources

counters, cubes, Shape song
Teacher's Guide page 16

Different: what does not belong?

Activity instructions

Look at each row of pictures. Put an X on the picture that does not belong.

Resources

counters, cubes
Teacher's Guide page 16

Taller

Activity instructions
Talk about the pairs of pictures. Circle the taller thing in each pair.

Resources
cubes
Teacher's Guide page 11

9

Shorter

Activity instructions

Talk about the pairs of objects. Colour the shorter object. How did you know that was the shorter one?

Resources

cubes
Teacher's Guide page 12

2 two

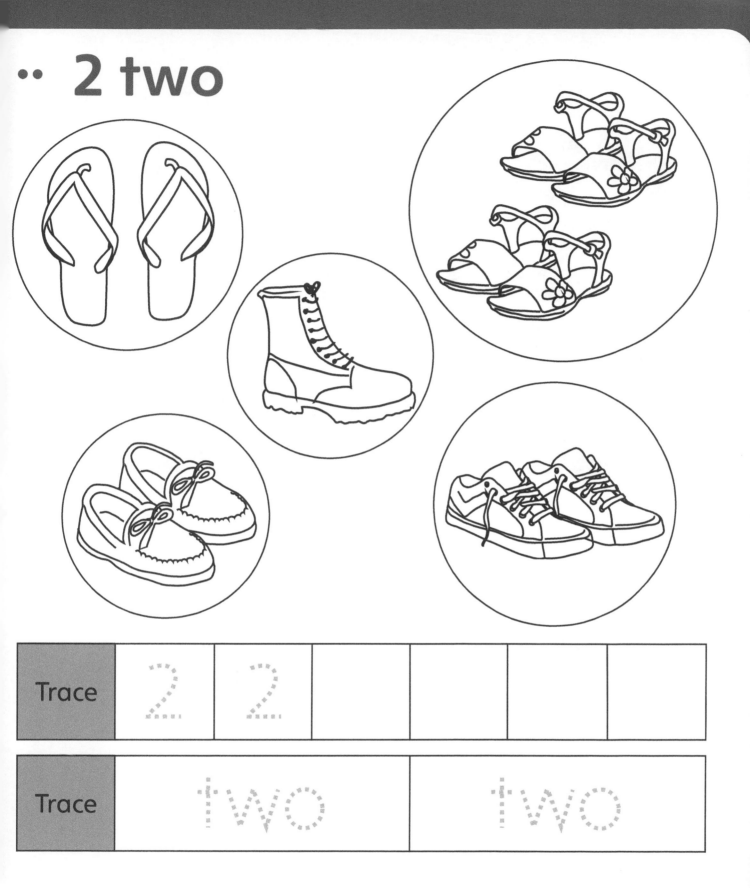

| Trace | 2 | 2 | | | | |

| Trace | two | | two |

Activity instructions
Count and colour the sets of two. Trace and write the numeral and number name.

Resources
flashcards, counters, songs: Number formations; One, two, buckle my shoe; Little Fingers
Teacher's Guide pages 6 and 8

Make them the same

Activity instructions

Look at the first picture and compare it to the second picture.
Talk about what is different. Use your pencil or crayon to make the
second picture look like the first picture.

Resources

counters, cubes
Teacher's Guide page 16

Shorter

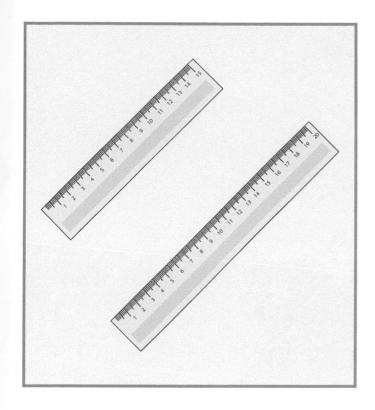

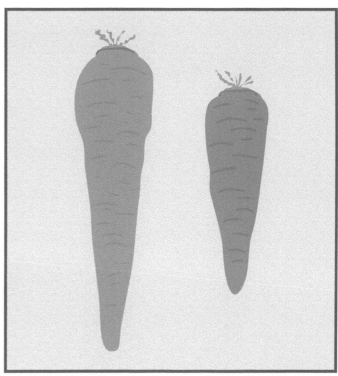

Activity instructions

Talk about the pairs of objects. Colour the shorter object. How do you know it's shorter?

Resources

cubes
Teacher's Guide page 13

Long

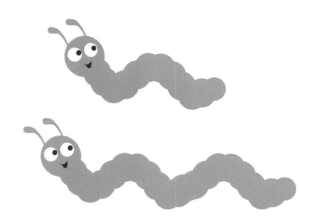

Activity instructions

Top: Draw a longer rope.
Middle: Draw a long tail on the cat.
Bottom: Circle the long caterpillar.

Resources

cubes
Teacher's Guide page 13

Holds more

Activity instructions

Talk about the pairs of containers. Put an X on the container that holds more.

Resources

Teacher's Guide page 15

15

Holds less

Activity instructions

Talk about the pairs of containers. Draw a circle around the container that holds less.

Resources

Teacher's Guide page 18

Full / empty

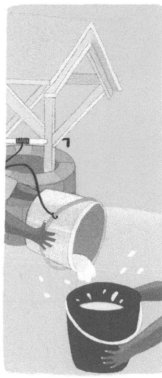

Activity instructions

Talk about the different pictures of Jack and Jill. Circle the full bucket. Put an X on the half full bucket. What are the other buckets that are left?

Resources

story: Jack and Jill
Teacher's Guide page 19

⠃ 3 three

Trace	3	3				

Trace	three	three

Activity instructions

Count and colour the 3 houses for the 3 pigs. Trace and write the numeral and number name.

Resources

story: The three little pigs, flashcards, counters, songs: One, two, buckle my shoe; Little fingers

Shapes: triangles and circles

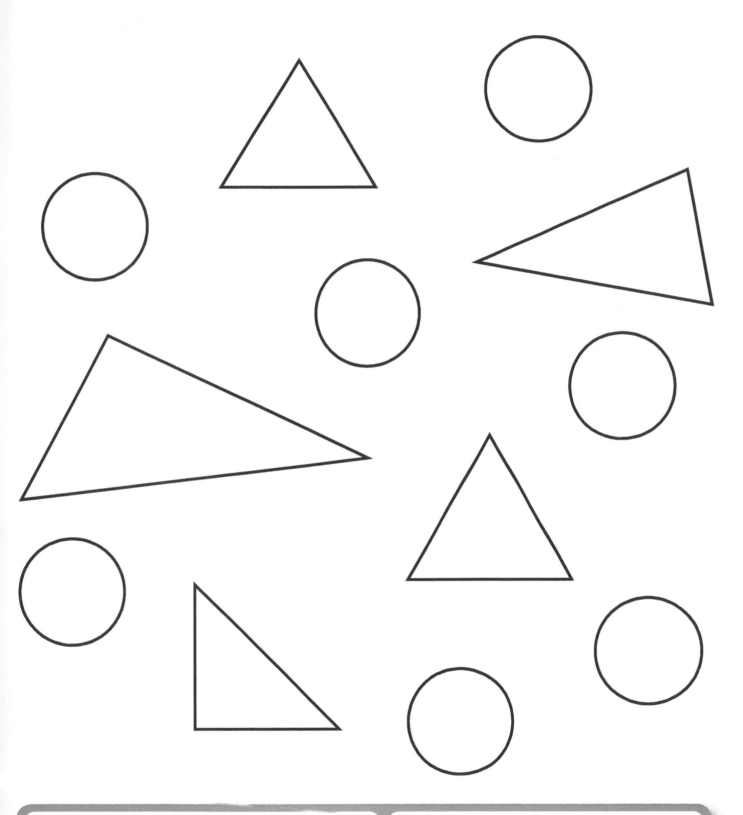

Activity instructions

Colour the triangles red.
Colour the circles green.

Resources

songs: Colours; Shape song
Teacher's Guide page 10

Day and night

Activity instructions

Look at the pictures of Jason. Talk about the different times of day in each picture. Is it morning, afternoon or night time? How do you know?

Resources

Teacher's Guide page 27

Day and night

Activity instructions

Draw lines to match the day activities to the Sun and the night activities to the Moon. Colour the Sun and the Moon.

Resources

Teacher's Guide page 27

Ordering

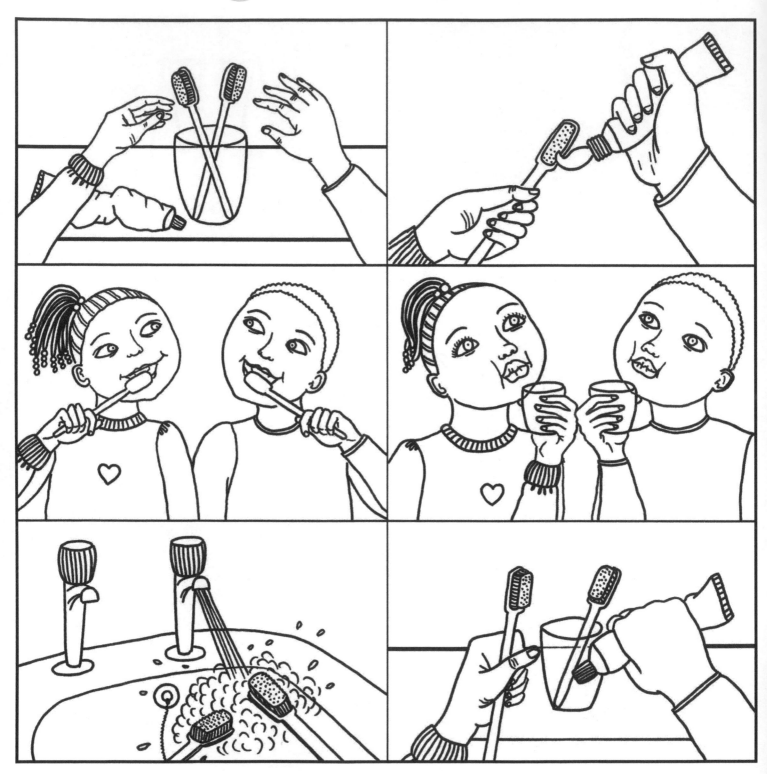

Activity instructions

Talk about what is happening in each picture. Think about the order of the pictures. Can you rinse your mouth out before you brush your teeth?

Resources

Teacher's Guide page 22

Patterns

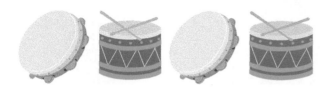

Activity instructions

Look at the patterns on the left of each line. Circle the picture on the right that comes next.

Resources

counters, cubes
Teacher's Guide page 24

23

One-to-one correspondence

Activity instructions

Help each parrot and dog find a home.
Top: Draw a line from the dog to the house with the same colour as
its collar.
Bottom: Draw a line from the parrot to the cage with the same
colour as its body.

Resources

parrot puppet, story: Goldilocks
Teacher's Guide page 5

One-to-one correspondence

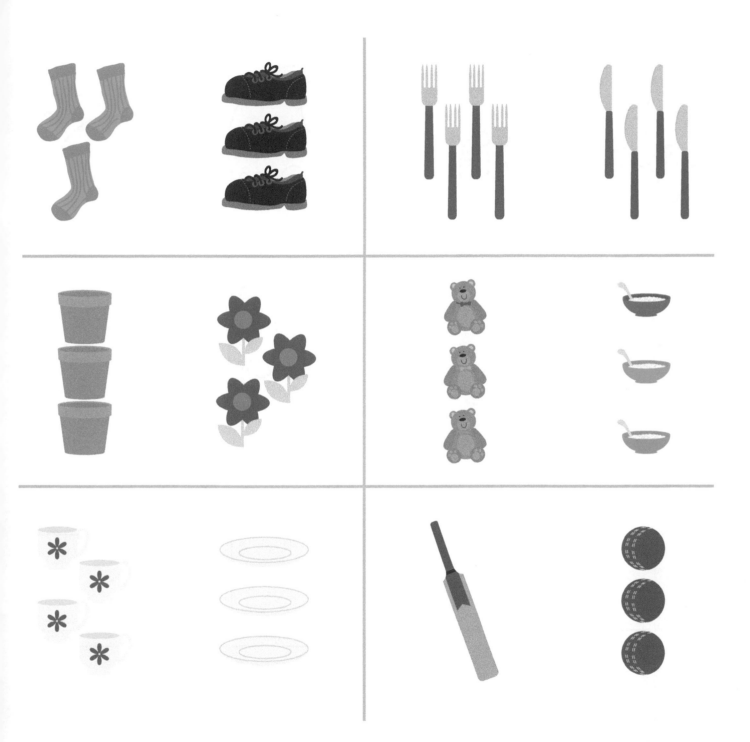

Activity instructions

Look at the things that go together, do they all have their own partner? Draw lines to help you answer.

Resources

Teacher's Guide page 5

25

∷ 4 four

Trace	4	4				

Trace	four	four

Shapes: squares and rectangles

Activity instructions

Colour the squares yellow.
Colour the rectangles blue.

Resources

songs: Colours; Shape song
Teacher's Guide pages 10 and 11

Positions: over, on, under

Activity instructions

Talk about the picture and the positions of the objects. Circle the object that is above the table. Put an X on the objects that are on the table. What is under the table?

Resources

parrot, song: Wheels on the bus
Teacher's Guide page 21

Positions: above and below

Activity instructions

Talk about the picture. Circle 2 things that are above the bridge.
Draw an X on 2 things that are below the bridge.

Resources

parrot
Teacher's Guide page 21

29

Looking back

Activity instructions

Colour 1 square yellow. Colour 2 triangles blue. Colour 2 rectangles red. Colour all of the circles green.

Resources

flashcards, counters, cubes, tangrams, shape counters, Colourful Shapes game, songs: My hat it has three corners; Shapes song Teacher's Guide page 10

Positions: top, middle, bottom

Activity instructions

Circle the object which is at the top. Colour the object which is at the bottom.

Resources

parrot puppet
Teacher's Guide page 21

Top, middle, bottom

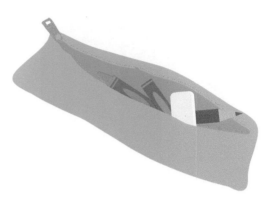

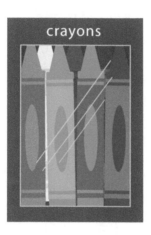
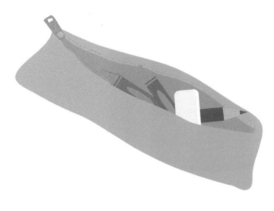

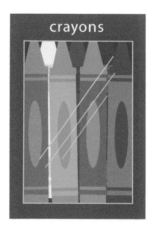
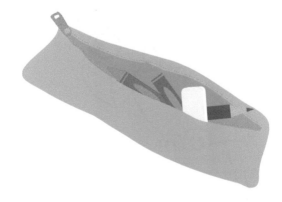

Activity instructions

Look at the picture of a shelf with a top, middle and bottom row.
Put an X on the crayons in the top row. Put an X on the ruler in the
middle row. Put an X on the pencil case in the bottom row.

Resources

parrot puppet
Teacher's Guide page 21

Looking back

Activity instructions

Colour the top rectangle red. Colour the middle rectangle yellow.
Colour the bottom rectangle blue.

Resources

Colourful Shapes game, songs: Colours; Shapes song
Teacher's Guide pages 10 and 21

Looking back

	1 one
 	3 three
	4 four

Activity instructions

Top: Draw 1 flower.
Middle: Draw 1 more car to make three.
Bottom: Draw more boats to make 4.
Trace the numerals and number names.

Resources

flashcards, counters, song: One, two, buckle my shoe
Teacher's Guide pages 8 and 9

5 five

Trace	5	5				

Trace	five	five

35

Making 5

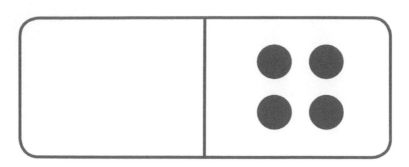

___ and 4 make ____

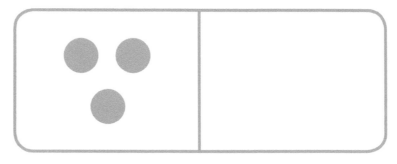

3 and ____ make ____

___ and ____ make 5

Activity instructions
Draw more to make 5. Fill in the blanks with the correct numerals.

Resources
frieze, flashcards, cubes
Teacher's Guide page 26

Counting

2

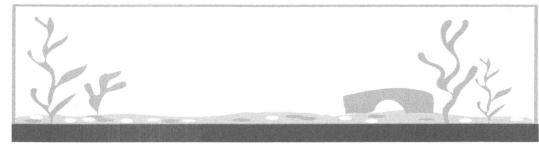

4

1

5

3

Activity instructions

Draw the number of fish in each tank to match the numeral shown next to each tank. You can also put the same number of counters in each tank.

Resources

flashcards, counters, song: Number formations
Teacher's Guide pages 8 and 9

Counting

one ★	two ★★	three ★★★	four ★★★★	five ★★★★★

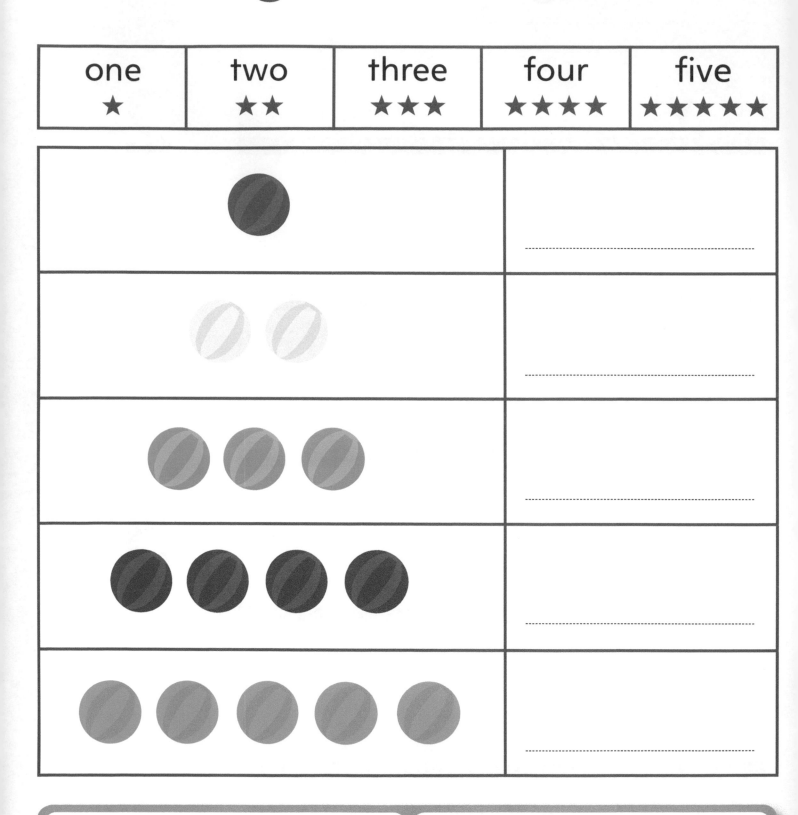

Activity instructions
Count the number of balls in each row. Write the matching numeral or number name for each set on the line.

Resources
frieze, flashcards, counters, cubes
Teacher's Guide page 8

A day at the beach

Activity instructions

Look at picture 1 and picture 2. Find and circle the things that are different in picture 2.

Resources

Teacher's Guide page 16

Sets

Activity instructions

Look at the picture. Discuss the number of objects in each group or set. Put a circle around each set of things that are the same. Draw a line to match the groups that have the same number of objects.

Resources

Teacher's Guide page 6

Looking back 👁 👁

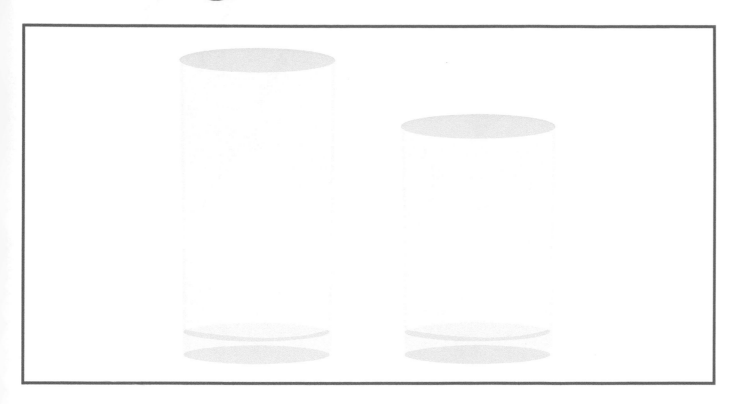

Activity instructions

Top: Circle the glass that holds less.
Bottom: Draw your own containers showing one holding more than the other.

Resources

Teacher's Guide page 18

6 six

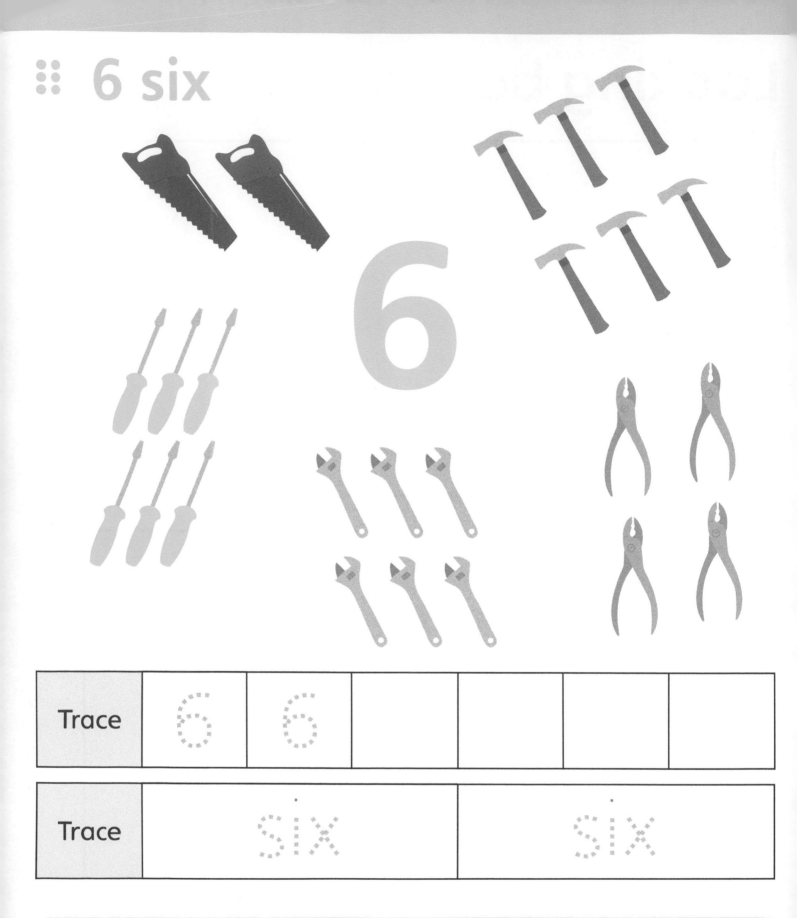

Trace	6	6				

Trace	six	six

Activity instructions

Top: Draw a line from the sets of 6 tools to the numeral 6 in the middle.
Bottom: Trace and write the numeral and number name.

Resources

frieze, flashcards, songs: Number formations; One two buckle my shoe
Teacher's Guide pages 8 and 9

42

Making 6

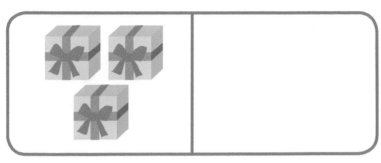

___3___ and _____ make __6__

____ and ____ make ____

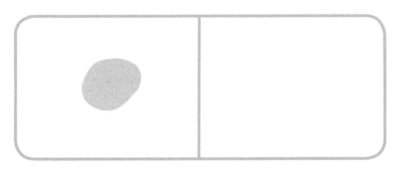

____ and ____ make ____

Activity instructions
Draw more to make 6. Fill in the blanks with the correct numerals.

Resources
counters, flashcards
Teacher's Guide page 26

43

Number patterns

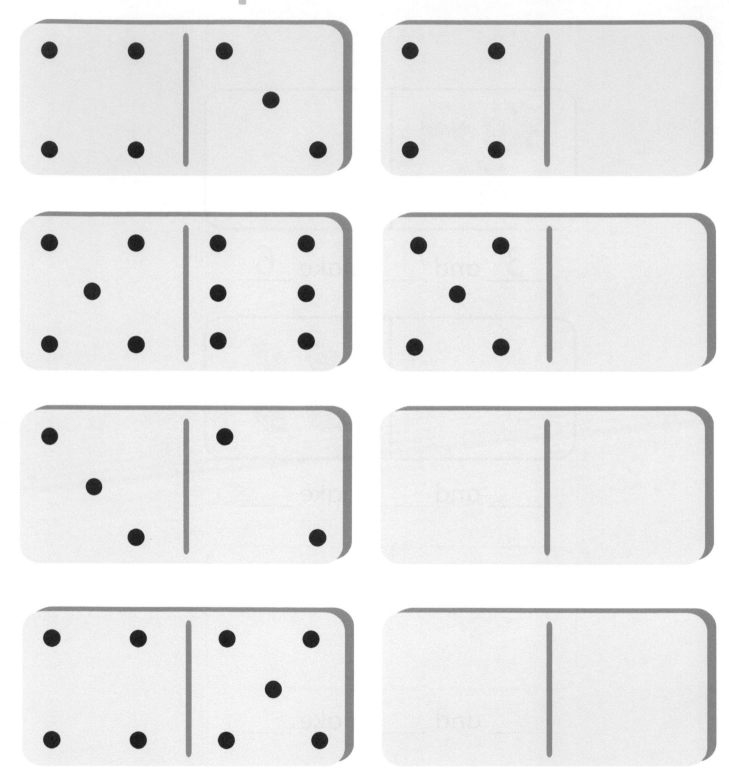

Activity instructions

Look at the first domino in each row. Draw the number of dots on the second domino to complete the pattern.

Resources

cubes, dice, flashcards, frieze
Teacher's Guide page 24

Number names

o	n	e	f	o	u	r
f	i	v	e	r	a	o
o	n	s	e	v	e	n
b	e	i	g	h	t	x
s	u	x	n	c	w	t
t	n	n	f	o	o	e
t	h	r	e	e	i	n

- one
- two
- three
- four
- five
- six

Activity instructions

Look at the number names above. Find them in the word search box and circle them. You can tick the words on the bottom when you find them to help you.

Resources

songs: Numbers; One, two, buckle my shoe
Teacher's Guide page 9

45

Picture graph

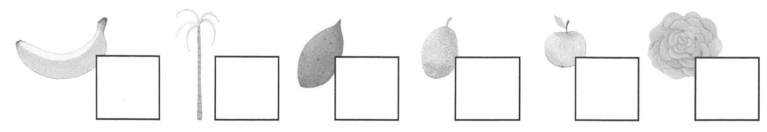

Activity instructions

Help Miss Pat count the fruit and vegetables on her stall. Count the objects and write the number in the box with that object's picture.

Resources

counters, flashcards, frieze
Teacher's Guide page 20

Looking back

one	▢	
three		
six		
five		
four		
two		

Looking back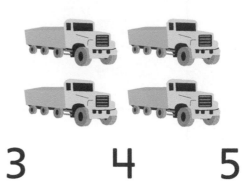

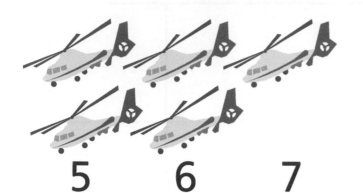

5   6   7   3   4   5

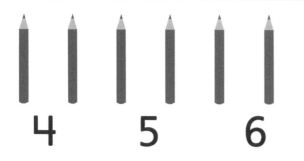

4   5   6   6   7   8

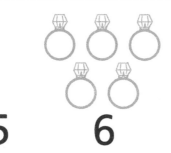

5   6   7   1   2   3

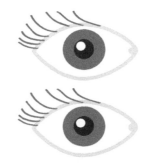

1
2
3

1
2
3

Activity instructions
Count the objects in each box and circle the correct answer.

Resources
frieze, flashcards
Teacher's Guide pages 8 and 9

Number recognition

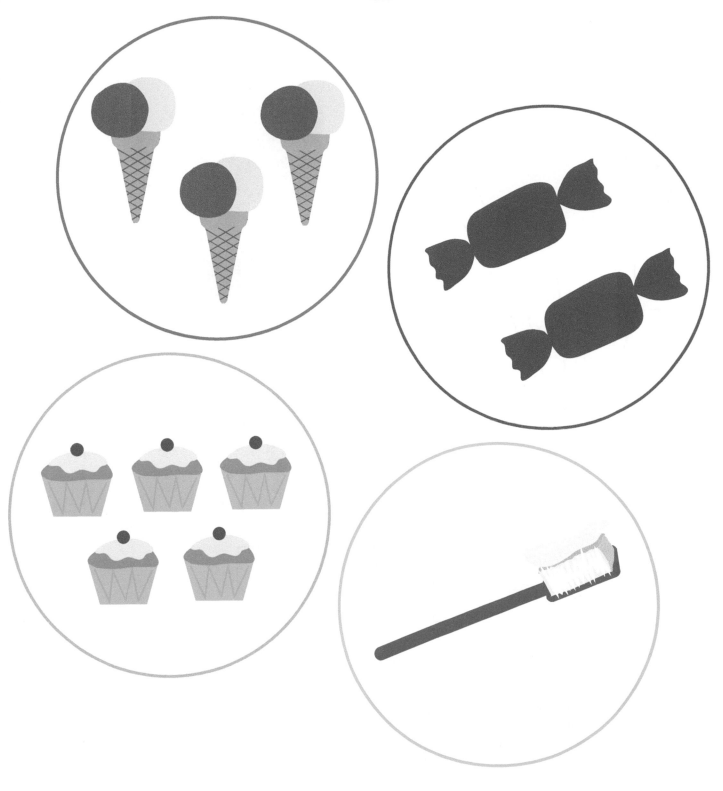

Activity instructions

Roll the dice. Put the corresponding number of counters on the group with the corresponding number of objects. Which 2 numbers are missing?

Resources

dice, counters, cubes
Teacher's Guide pages 5, 8 and 9

It's a party – let's count

Activity instructions

Count the different groups of pictures. Write the answer in each box.

Resources

counters, frieze, flashcards
Teacher's Guide pages 6, 8 and 9

Counting

Count and circle the correct answer.

1   2   3

4   5   6

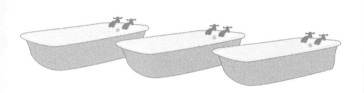

1   2   3

1   2   3

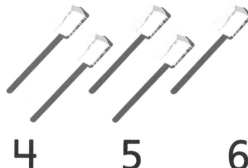

4   5   6

1   2   3

Activity instructions

Count and circle the correct answer in each box.

Resources

counters, flashcards
Teacher's Guide pages 6, 8 and 9

Looking back

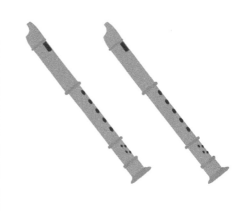

Half

Activity instructions

Talk about the pictures. Draw a line from the picture of the whole fruit to the picture of the half fruit.

Resources

Teacher's Guide page 27

Half

Activity instructions

Talk about the pictures. Colour one half of each picture.

Resources

flashcards, counters
Teacher's Guide page 27

Number recognition

5 five

2 two

6 six

3 three

1 one

Activity instructions

Roll the dice then put the same number of counters on the numeral.
Which number on the dice is missing from this page?

Resources

dice, counters, cubes
Teacher's Guide pages 8 and 9

55

Left and right

Left	Right

Activity instructions

Look at the pictures on the left and right. Colour the leaf on the right green. Colour the monkey on the left brown. Colour the radio on the right black.

Resources

song: Hokey cokey
Teacher's Guide page 27

Left and right

Left

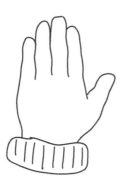

Right

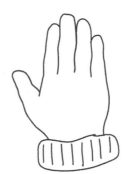

7 seven

Trace	7	7				

Trace	seven	seven

Activity instructions

Top: Count and circle the sets of seven bees.
Bottom: Trace and write the numeral and number name.

Resources

frieze, flashcards, counters
song: Number formations
Teacher's Guide pages 8 and 9

58

Making 7

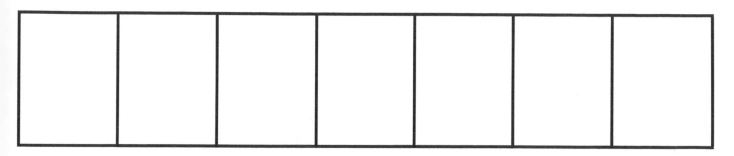

5 and 2 make ____

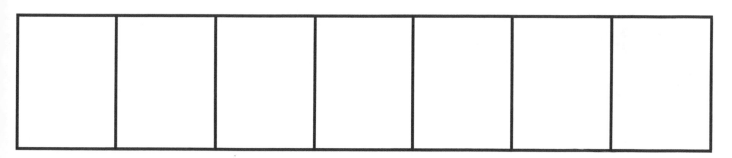

3 and 4 make ____

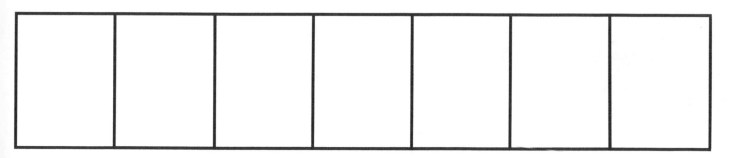

I and ____ make ____

Which number comes next?

1 2 3 4 5 6 7 8 9 10 11 12

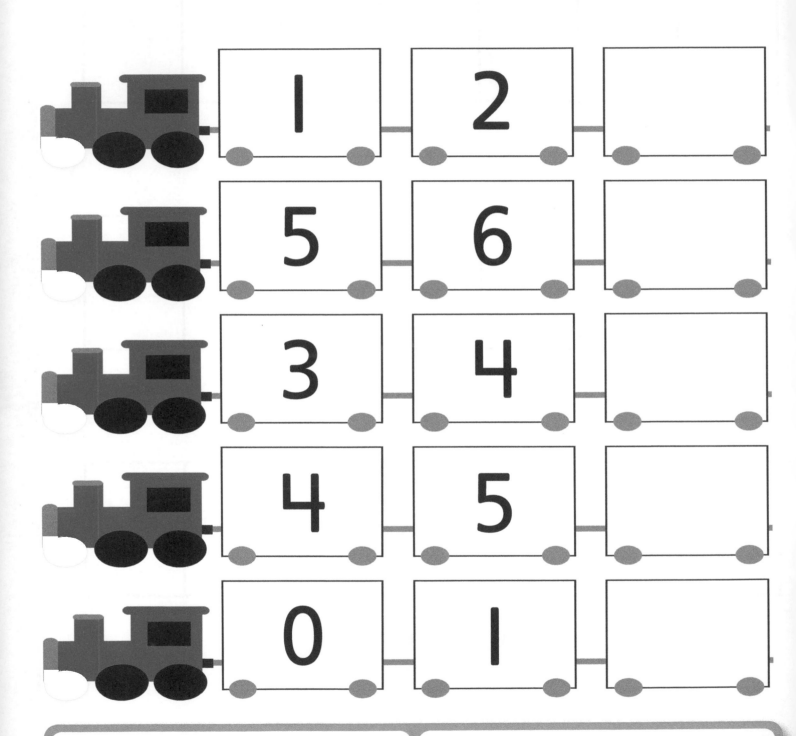

Activity instructions

Write the numeral that comes next on each train's car. Use the number line at the top to help you.

Resources

frieze, songs: One, two, buckle my shoe; Little fingers; Numbers Teacher's Guide page 26

Ordinals

Activity instructions

Listen to the story of the Tortoise and the Hare. Talk about who won the race. Top: Colour the animal that came first in the story. Bottom: Colour the first parrot red, the second parrot blue, the third parrot orange and the fourth parrot green.

Resources

parrot puppet, story: The tortoise and the hare
Teacher's Guide page 23

Ordinals

Activity instructions

Top: Look at the picture of the horses racing. Circle the horse that is in third position.
Middle: Circle the first rooster.
Bottom: Circle the fifth flamingo.

Resources

frieze, flashcards
Teacher's Guide page 23

Less than

Activity instructions

Top: Circle the set of cars that has less.
Bottom: Circle the set of bikes that has less.

Resources

counters, cubes
Teacher's Guide page 14

More than

Activity instructions

Count the sets of objects on the left and right side. Circle the set that has more in each row.

Resources

counters, cubes
Teacher's Guide page 14

8 eight

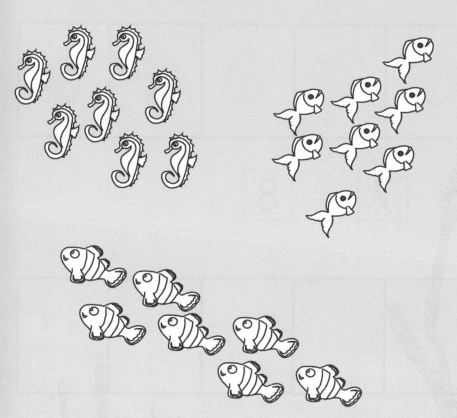

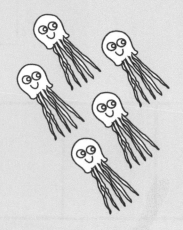

Trace	8	8				

Trace	eight	eight

Activity instructions

Top: Count and colour the sets of 8 sea creatures.
Bottom: Trace and write the numeral and number name.

Resources

frieze, flashcards, counters, songs: Number formations;
One, two, buckle my shoe
Teacher's Guide pages 6, 8 and 9

Making 8

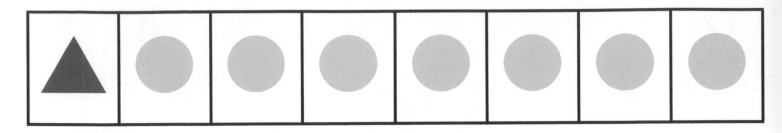

$$\boxed{1} + \boxed{7} = \boxed{8}$$

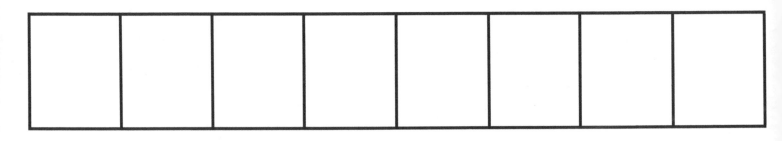

$$\boxed{} + \boxed{} = \boxed{}$$

$$\boxed{} + \boxed{} = \boxed{}$$

Activity instructions

Write the number sentence for each line. Use the example to help.

Resources

counters, cubes
Teacher's Guide page 26

Making 8

3 and 5	4 and 4	4 and 6
8 and 0	Making 8	3 and 3
3 and 1	3 and 2	5 and 2

Activity instructions
Find and colour the boxes that make 8. Use counters or strokes to help solve the problems.

Resources
counters, cubes
Teacher's Guide page 26

Zero

Trace	◌	◌				

Trace	zero	zero

Activity instructions

Top: Look at the pictures on the page. How many black balloons do you see? How many pink teddy bears do you see? How many birds do you see?
Bottom: Trace the numeral and number name.

Resources

Teacher's Guide page 9

Looking back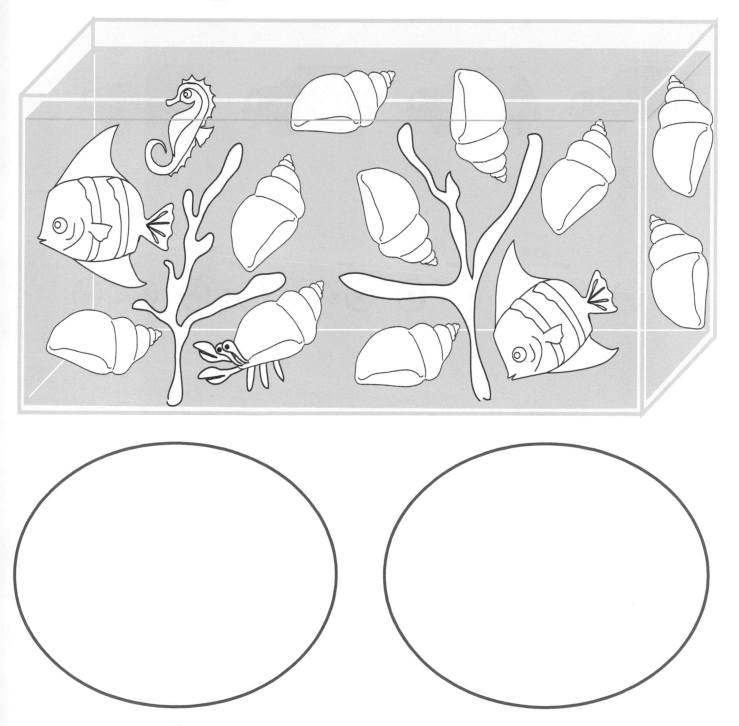

8

0

Activity instructions

Top: Colour 8 shells for Hermit crab.
Bottom: Draw objects to match the numerals.

Resources

frieze, flashcards, counters
Teacher's Guide page 8

One less

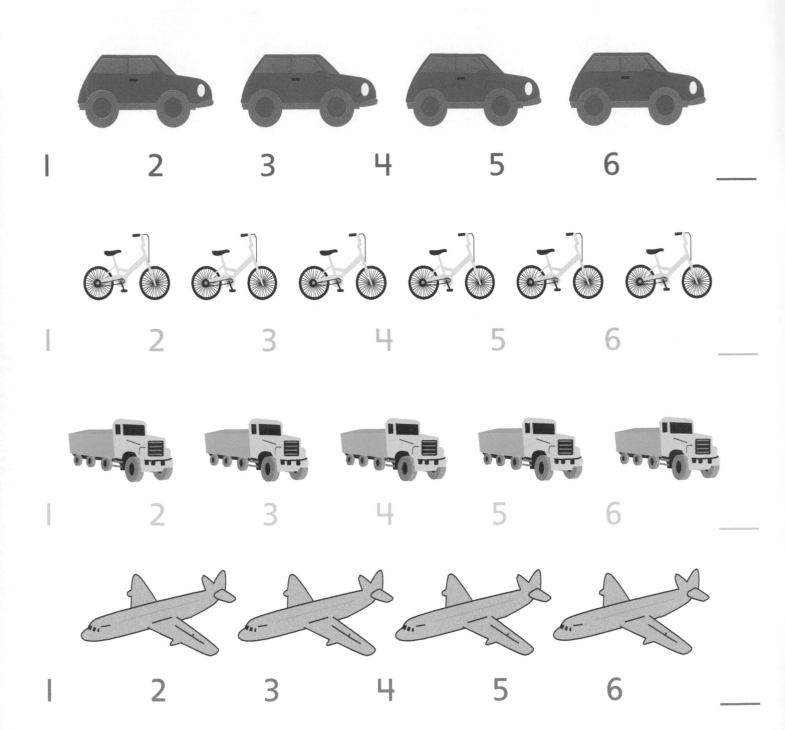

1   2   3   4   5   6   ___

1   2   3   4   5   6   ___

1   2   3   4   5   6   ___

1   2   3   4   5   6   ___

Activity instructions

Count the number of objects in each row. Then, cross one object off and say how many are left. Circle the number on the number line that represents the new total in this group.

Resources

frieze, flashcards, counters
Teacher's Guide page 30

Looking back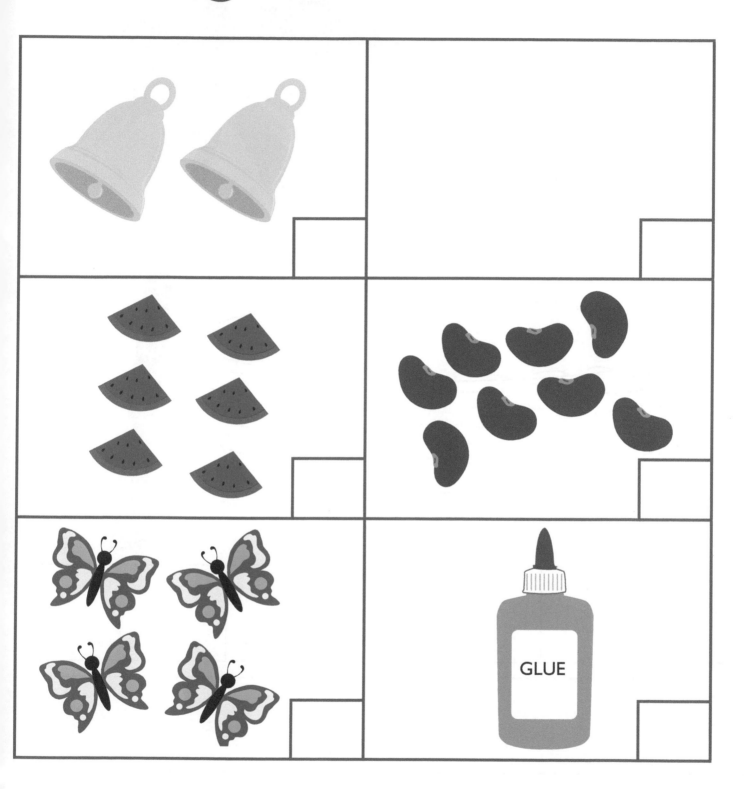

Activity instructions

Count the objects in each set. Write the answer in the box provided.

Resources

counters
Teacher's Guide page 8

9 nine

Trace	9	9				

Trace	nine	nine

Activity instructions

Top: What is your favourite fruit? Draw and colour 9 of them for the parrot to eat.
Bottom: Trace and write the numeral and number name.

Resources

counters, parrot puppet, songs: Numbers; Number formations
Teacher's Guide pages 8 and 9

72

Making 9

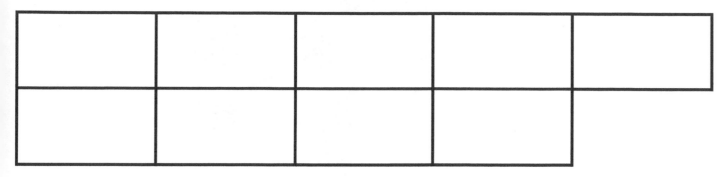

1 + 8 = ____

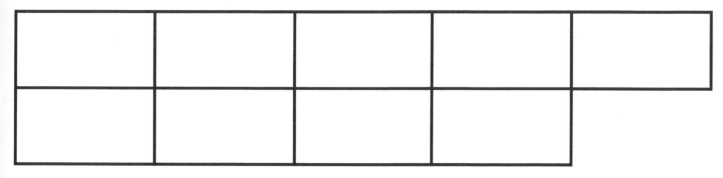

5 + ____ = 9

____ and ____ make ____

Activity instructions

Look at the number sentences. Use red and yellow crayons to colour the boxes and show 9 according to the number sentences. Fill
in the number sentences. Make your own combination of red and
yellow to make 9 for the last box.

Resources

cubes
Teacher's Guide page 26

Count my spots

Activity instructions

Count the number of spots on each animal and write the numeral in the box. Circle the two animals that have an equal number of spots.

Resources

frieze, flashcards, counters
Teacher's Guide pages 8 and 9

Looking back

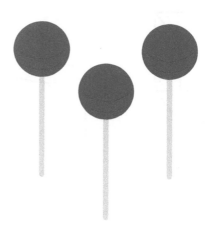

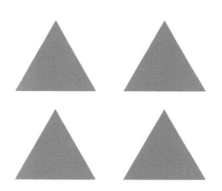

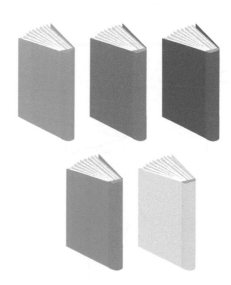

Activity instructions

Top: Draw pictures to show less.
Middle: Draw pictures to show less.
Bottom: Draw pictures to show more.

Resources

counters, flashcards
Teacher's Guide page 15

75

Looking back 👁 👁

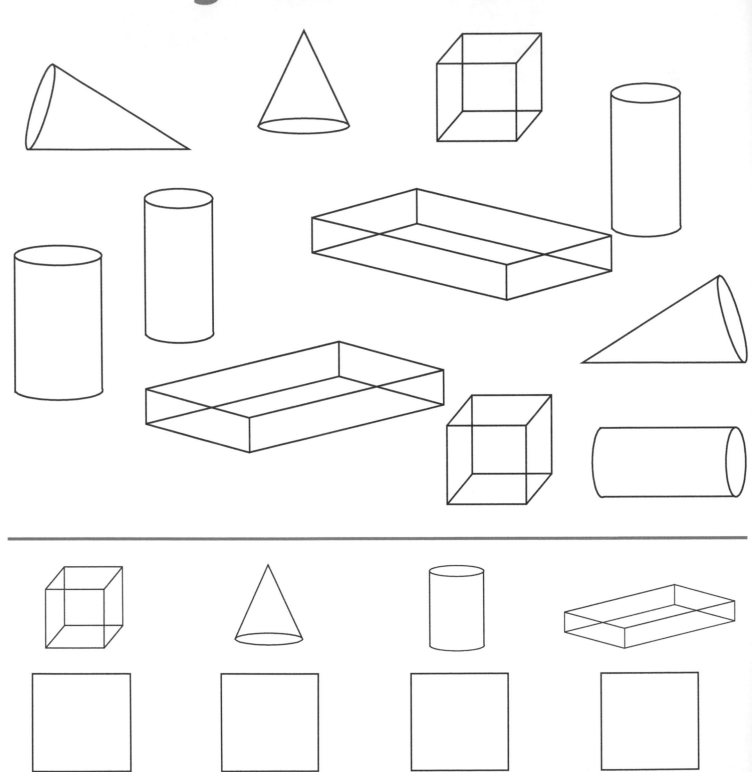

Activity instructions

What shapes can you see? Count the shapes and write the numerals in the boxes. Colour the shapes.

Resources

Colourful Shapes game, counters, song: Shapes Teacher's Guide page 10

76

Looking back

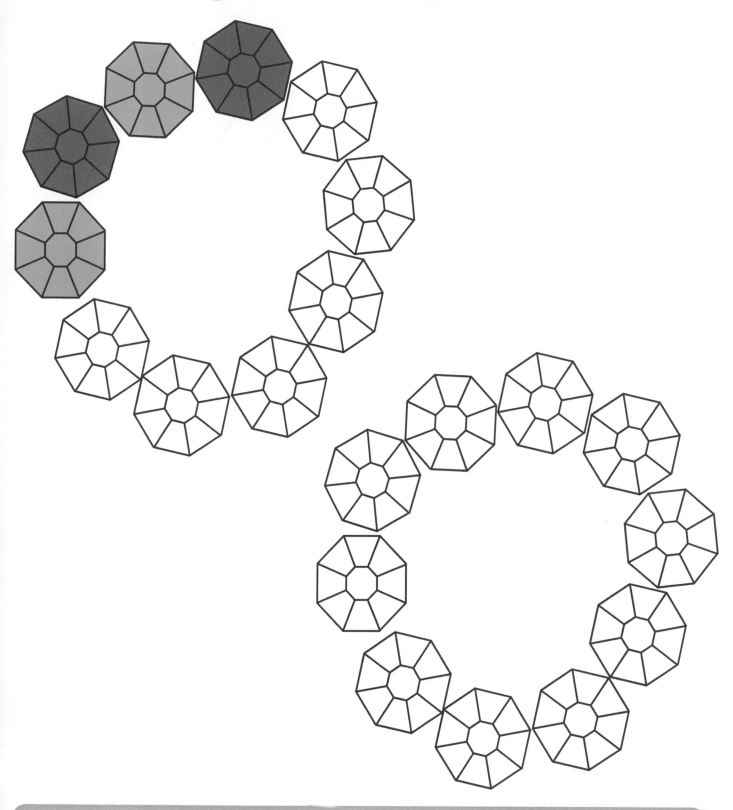

Activity instructions

Complete the colour pattern on the necklace at the top. Make your own pattern for the necklace on the bottom. Talk about the pattern you made.

Resources

cubes
Teacher's Guide page 24

10 ten

Trace	10 10				

Trace	ten ten

Activity instructions

Top: Colour 10 feathers on the parrot.
Bottom: Trace the numeral and number name.

Resources

parrot puppet, counters, flashcards
song: Ten green bottles
Teacher's Guide pages 8 and 9

Making 10

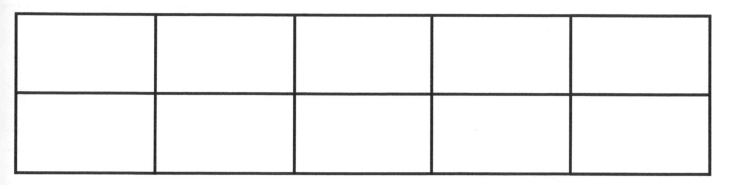

$$10 + \underline{\quad} = 0$$

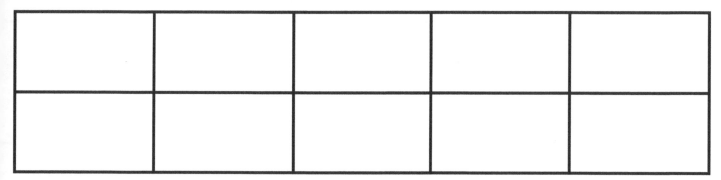

$$\underline{\quad} \text{ and } \underline{\quad} = 10$$

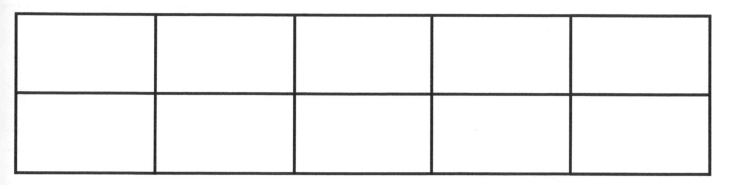

$$\underline{\quad} \text{ and } \underline{\quad} = \underline{\quad}$$

Activity instructions
Use yellow and green crayons to colour to show ten.
Write the number sentence for each problem.

Resources
counters, songs: Numbers; Ten green bottles
Teacher's Guide page 26

Counting

six	seven	eight	nine	ten
★ ★ ★ ★ ★ ★	★ ★ ★ ★ ★ ★ ★	★ ★ ★ ★ ★ ★ ★ ★	★ ★ ★ ★ ★ ★ ★ ★ ★	★ ★ ★ ★ ★ ★ ★ ★ ★ ★

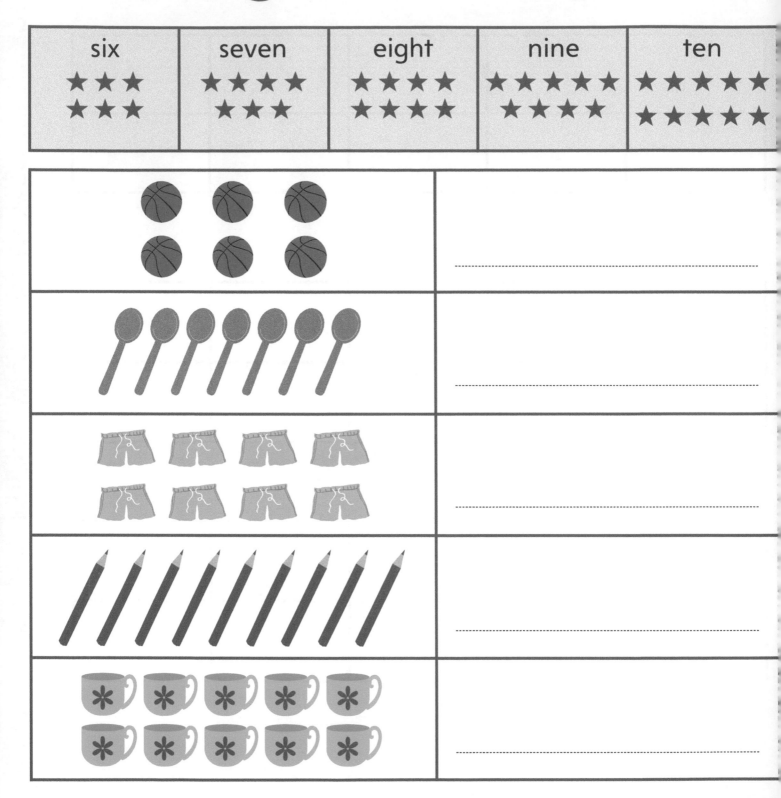

Activity instructions

Count the number of objects on each line. Write the numeral or number name that corresponds with each set of objects.

Resources

frieze, flashcards, counters, songs: Numbers; Number formations
Teacher's Guide page 8

Counting

Activity instructions

Write the numbers in order from 1-10 on the flowers. Colour all the flowers.

Resources

counters, flashcards
Teacher's Guide page 8

Sets: how many?

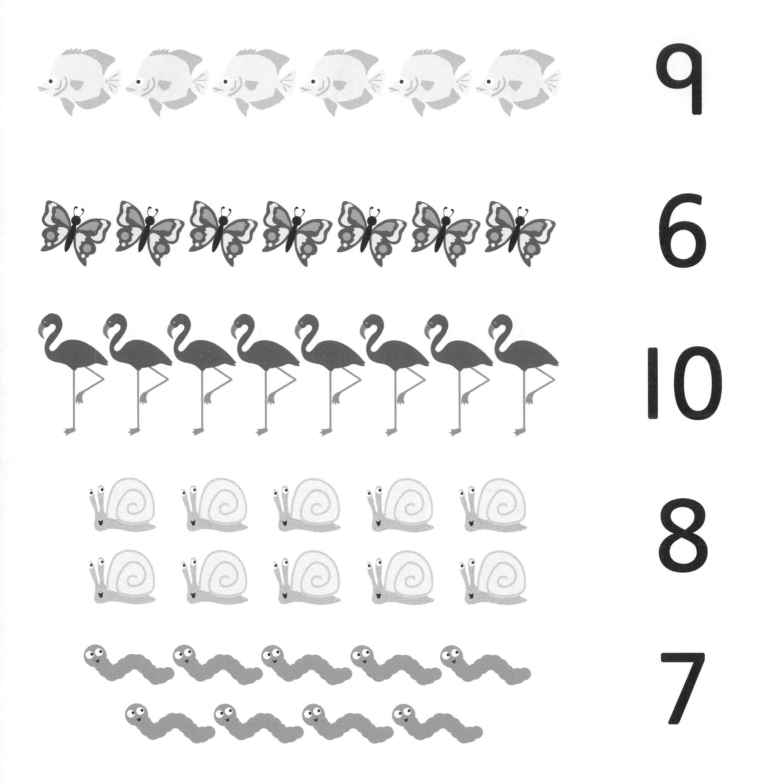

9

6

10

8

7

Activity instructions

Count the objects on each line. Draw a line from each set of objects to the corresponding numeral.

Resources

counters, flashcards
Teacher's Guide pages 8 and 9

How many?

Activity instructions

Count the dots on each ladybug. Write the numeral beside each ladybug.

Resources

counters, frieze, song: Numbers

Missing numerals

CALENDAR				JULY		
1	2	3	4	5	6	7
	9	10	11	12	13	14
15	16	17	18	19	20	21
22	23	24	25	26	27	28
29	30	31				

Graphs

_____   _____

Activity instructions

Count the number of cows at the top of the page and place counters to show how many on the graph below. Do the same for the horses. Write the corresponding numeral on the line. Are there more cows or horses?

Resources

counters
Teacher's Guide page 20

Graphs

5	4	3	2	1	0

Activity instructions

Use any colour you choose to colour the number of squares to represent the numeral on the bottom. What do you notice about the graph?

Resources

counters, The Squeeze game
Teacher's Guide page 20

Looking back 👁 👁

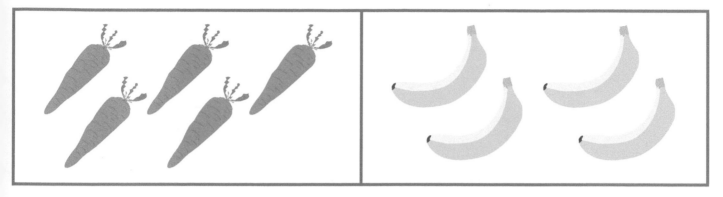

5 and 4 is __9__

2 and 5 is ____

4 and 6 is ____

Activity instructions

Draw pictures in the boxes to represent the number sentences

Resources

counters, cubes
Teacher's Guide page 26

Joining two sets

4 frogs   2 frogs

6 in all

____ fish   ____ fish

____ in all

____ sheep   ____ sheep

____ in all

Activity instructions

Look at the pictures on each line. Count the objects and complete the number sentences. Use the example at the top to help.

Resources

counters, cubes
Teacher's Guide page 28

Looking back

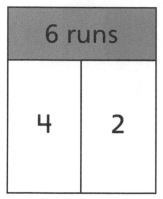

6 runs

4	2

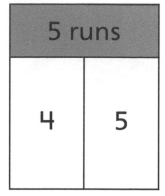

5 runs

4	5

10 runs

9	1

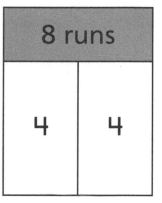

8 runs

4	4

9 runs

8	1

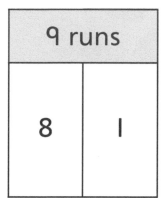

7 runs

0	1

Activity instructions

Each inning has a different score. Circle the boxes whose two innings add up to the same number of runs on each scoreboard above. Use strokes or counters to help.

Resources

counters, cubes
Teacher's Guide page 28

Missing numbers

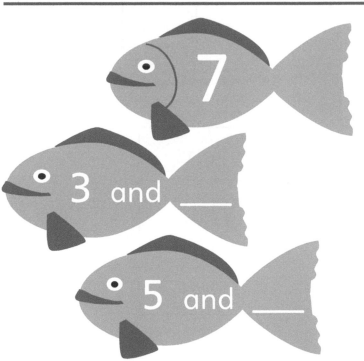

Activity instructions
Look at the sum on the first fish. Write the missing number on each fish to make this number. Use strokes or counters to help.

Resources
counters, cubes
Teacher's Guide page 28

11 eleven

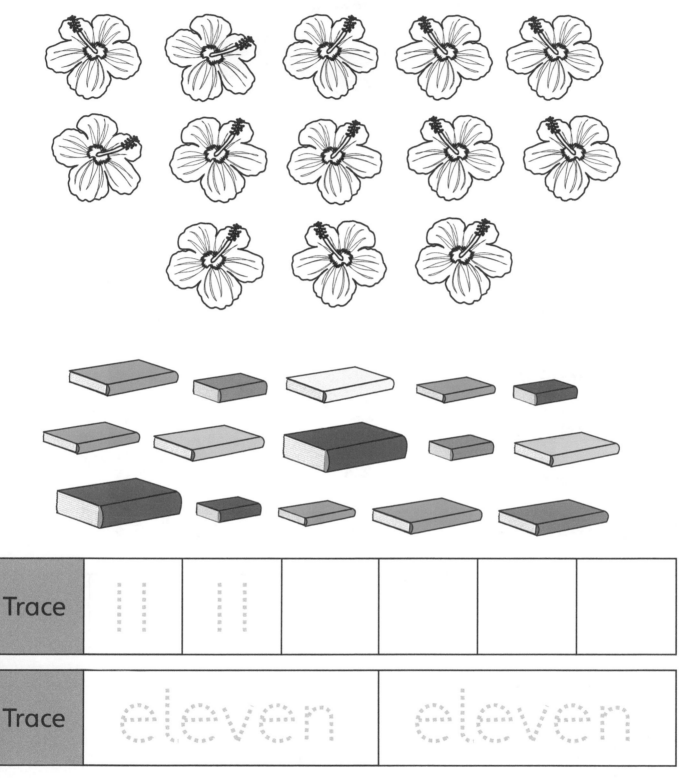

Trace	11	11				

Trace	eleven	eleven

Activity instructions
Top: Colour 11 flowers.
Middle: Circle 11 books.
Bottom: Trace and write the numeral and number name.

Resources
frieze, flashcards, counters, songs: Numbers; Number race
Teacher's Guide pages 8 and 9

12 twelve

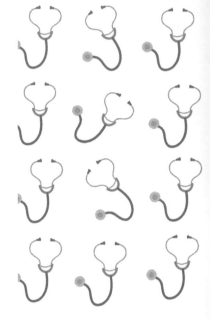

12

Trace	12 12				

Trace	twelve	twelve

Activity instructions

Draw a line from the sets with 12 objects to the 12 in the middle.
Trace and write the numeral and number name.

Resources

frieze, flashcards, counters, songs: Numbers; Number race
Teacher's Guide pages 6, 8 and 9

Counting 11 and 12

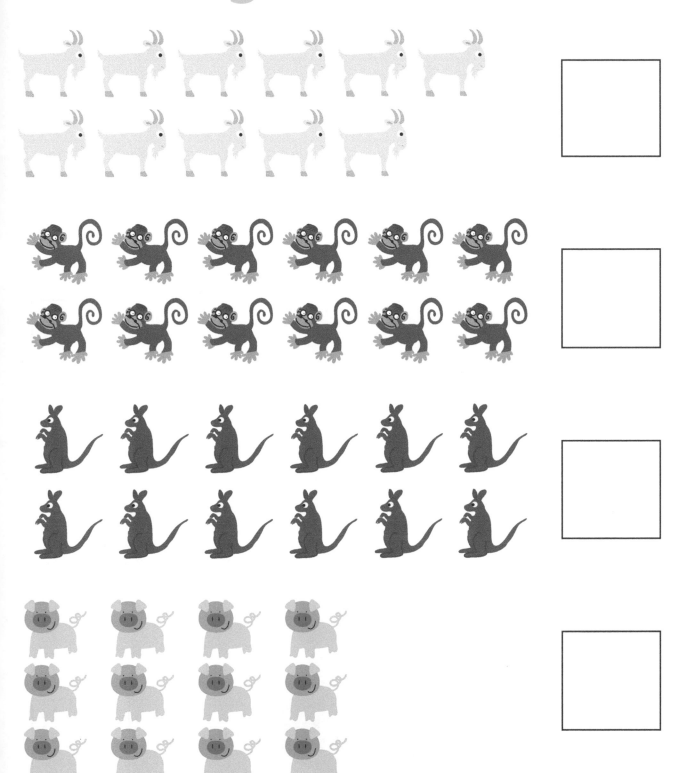

Activity instructions

Count the sets of animals on each line. Write the same numeral in the boxes.

Resources

flashcards, counters, cubes
Teacher's Guide pages 6, 8 and 9

93

Looking back

Activity instructions

Top: Draw eleven toys. Write the numeral 11 in the box.
Bottom: Draw 12 flowers in the flower pot.

Resources

counters, cubes, songs: Numbers; Number race
Teacher's Guide page 8

Looking back 👁️ 👁️

12   5   11

8   9   10

5   7   2

6   3   10

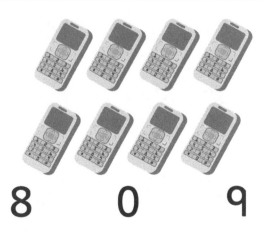

8   0   9

4   11   7

Activity instructions

Count and circle the correct answer.

Resources

frieze, cubes
Teacher's Guide pages 8 and 9

Counting backwards

1 2 3 4 5 6 7 8 9 10

Activity instructions

The frog jumped from waterlily to waterlily. Draw the frog's path
from 10 to 1 in order. Use the number line above to help you.

Resources

frieze, song: Ten green bottles
Teacher's Guide page 31

96

Vertical addition

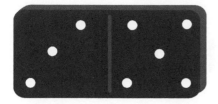

3 + 5 = 8

$$
\begin{array}{r}
3 \\
+\ 5 \\
\hline
8
\end{array}
$$

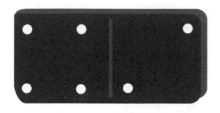

☐ + ☐ = ☐

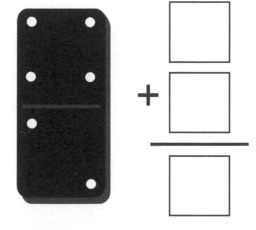

☐
+ ☐
―――
☐

☐ + ☐ = ☐

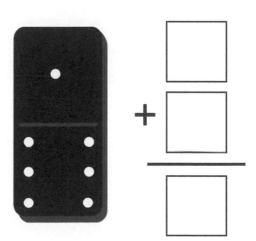

☐
+ ☐
―――
☐

Activity instructions
Write the numeral to match each dot. Write the sum.

Resources
counters, cubes
Teacher's Guide page 28

Before

1 2 3 4 5 6 7 8 9 10 11 12

Activity instructions

Write the numeral that comes before the given number in each row.
Use the number line at the top to help you.

Resources

frieze,
Teacher's Guide page 29

After

1 2 3 4 5 6 7 8 9 10 11 12

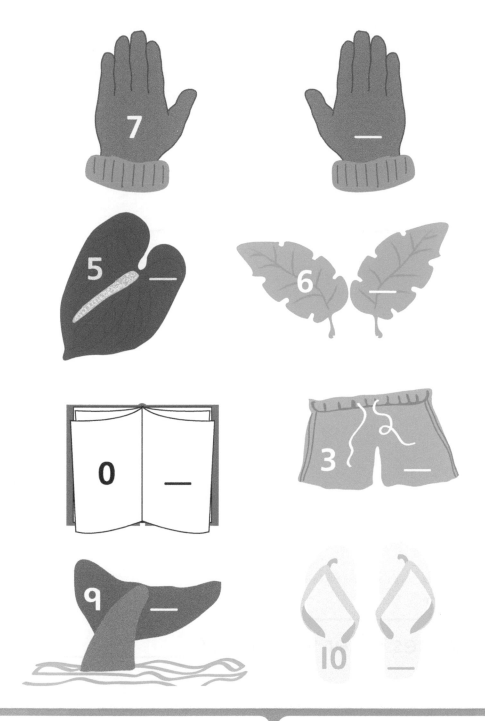

7 ___

5 ___   6 ___

0 ___   3 ___

9 ___   10 ___

Activity instructions
Write the numeral that comes after the numeral in each set. Use the number line at the top to help you.

Resources
counters, cubes, song: One, two, buckle my shoe
Teacher's Guide page 29

Before and after

0 1 2 3 4 5 6 7 8 9 10

_____ 5 _____

_____ 1 _____

_____ 9 _____

_____ 7 _____

What is your age? _____
What number comes after your age? _____

Looking back

_____

Activity instructions

Top: Look at the pictures. Circle the one that holds more.
Middle: Look at the shapes, colour the rectangle blue.
Bottom: Count the number of butterflies and write the answer on the line.

Resources

frieze, Colourful Shapes game, songs: Shapes, One, two, buckle my shoe
Teacher's Guide pages 10 and 11

More than/less than

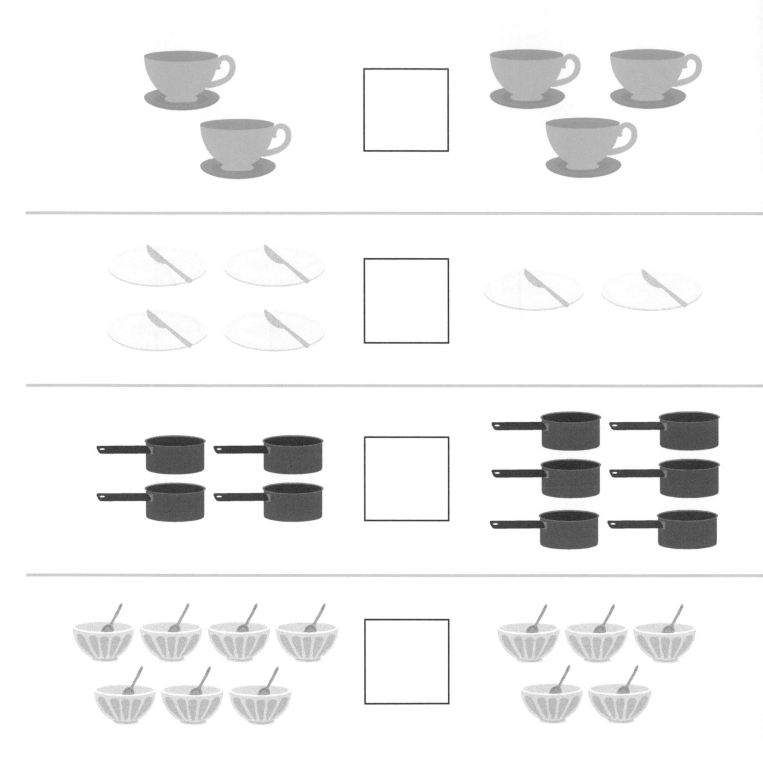

Activity instructions

Look at the items on each row, Does the left hand have more or less than the right side? Put the correct sign in the box on each row.

Resources

counters
Teacher's Guide pages 14 and 15

Take away

_4_ take away 2 is _2_

_5_ take away 1 is _____

_____ take away 4 is _____

Activity instructions

Count how many objects are in each set. Write it down on the first line under the set. Put an x on the number of objects we have to take away. Count the number of objects that are left and write it on the second line.

Resources

counters
Teacher's Guide page 30

Extension: colour by numbers

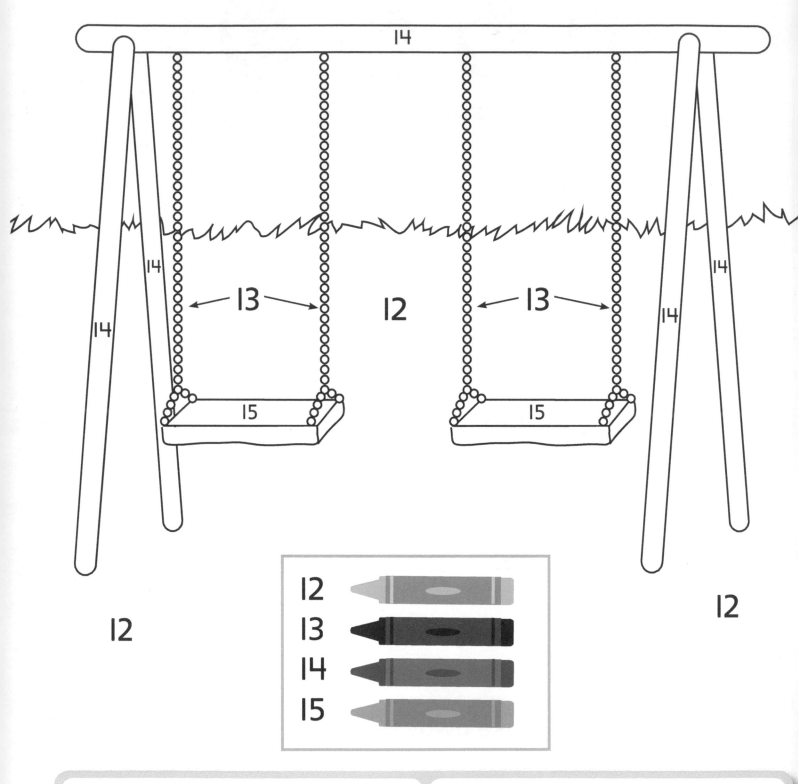

Activity instructions

Use the code to help you to colour the picture.

Resources

song: Colours
Teacher's Guide page 9